Animal Fights

Cataloging Information

Ham, Catherine.
Animal fights/Catherine Ham
32 p. : col. ill. ; 20 cm.
Includes index (p.).
Summary: Uses verse and photographs to explore sleep and sleep-related behavior in animals. Includes a range of taxa, including mammals, insects, birds, reptiles, and fish.
LC: QL758.5 .K55
Dewey: 591.56/6 21
ISBN-13: 978-0-9832014-0-3 (alk. paper)
Fights —Juvenile literature

Cover Design: Stewart Pack
Art Director: Celia Naranjo
Copy Editor: Tina Miller
Photo Research: Dawn Cusick

10 9 8 7 6 5 4 3 2 1

First edition

For Kristen . . .

Published by EarlyLight Books, Inc.
1436 Dellwood Road
Waynesville, NC 28786

ISBN 13: 978-0-9832014-0-3

from: Greta Federspiel
2025

Animal Fights

Catherine Ham

WAYNESVILLE, NORTH CAROLINA, USA

Animal Fights

Squawking and squabbling
All kinds of animals
Bellowing, babbling
Biting and such

Lions and lizards
Egrets and elephants
Bee eaters, buffalo
Bickering much

KICK!

When a zebra gets into a fight
He uses his own special trick
His back legs are long
And amazingly strong
And deliver a powerful kick!

A tiger's not usually a fighter
And will often roll onto his back
If he thinks another male tiger
Is planning to launch an attack.
He may first try to make ugly faces
He may hiss, spit, moan, growl and roar
But if the other guy starts showing traces
Where it seems that he's likely to score
Then watch out!
For no doubt
He'll soon fetch him a clout
And knock him about
With his powerful front paws

And if the fight starts getting wilder
They'll both turn to using their claws
At the throat they will grab
With their paws they will stab
Until tight they can nab
With their jaws

HISS, SPIT ROAR!

GNAW!

The camel was born
Without any horn
The only way he can fight
Is to BITE!

BASH!

Have ever you seen such a curious sight
As two giraffes in a furious fight?
They wrestle and twist
With their long, long necks
They push and they pull
And try to crack heads
But when one gets it wrong...
Smack!
THWACK!
Oh me! oh my!
Bash!
CRASH!
Do you think they might cry?

ATTACK!

A bear -- big, small, white, brown or black
Can sometimes be very quick to attack.
If he thinks someone's stealing his food
He's likely to be in a very bad mood.
He has great nasty claws
On the ends of his paws
He can lash out and gash
He can bash out and slash
If he can hug his enemy terribly tight
Then he can horribly, dreadfully bite
With those razor-sharp teeth
Set so deep in his jaws.

SLASH & GASH!

JAM!

RAM!

Tusks are the elephants' weapon
And they'll wield them with all of their might
If a bull elephant
Is challenged to a fight....
If a bull elephant
Is fighting for the right
To win a lady elephant
He'll spread his ears out wide
And roaring, charge the other guy
Who'll have no place left to hide

Once they begin the battle
They push and shove up tight
They RAM their tusks together
They SLAM their tusks together
They JAM their tusks together....
Two tremendous elephants
In a most stupendous fight!

LOCKING HORNS

This strong and agile bearded goat
Quite fearless when it climbs
In Europe, Asia, Africa
Has been known since ancient times

These wild mountain goats are endangered
Their worst enemy, sadly, is Man
They graze in most perilous places
Eating all the plant matter they can

Ibex have horns curving backwards
Which are used to protect and to fight
When two males decide to lock horns
It's a simply incredible sight!

MONKEY FIGHTS

Monkeys live in a family group
A bunch of them together
Is known as a troop
Monkeys are noisy
Boisterous and loud
They chitter and they chatter
In a racket of a crowd
They twitter and they natter
As they scout about for food
But often when they find some
They get in a fighting mood

SCREECH!

Even when there's plenty
Quite enough to go around
They'll nab and grab
And shove and screech
They'll bark and whoop
And push and reach
They'll biff and bump
They'll grump, they'll thump
And sometimes they will quickly jump
Right into a fight

But even though monkeys are rowdy
They'd really far rather say "Howdy!"
Than fight
But of course they will bite
If they're under attack
What else can they do
If they have to fight back?

Rhinos are massive and lumpy
And really most horribly grumpy
They'll fight for their space
With that horn on their face
Making all those around them quite jumpy

A hippo is simply enormous
GINORMOUS
And very quick to pick a fight
He tosses his colossal head
And bellows with all of his might
He opens his gigantic mouth
And yikes! what a hideous sight
Those great...big...huge
And sharp front teeth...
You can almost die of fright

Oh please!
Please please please
Don't bite!

CRUSH!

Seals spend most of their lives in the water
And come very seldom ashore
Which is why we know little about them
But we're working on learning some more

They live by the hundreds
In groups very large
Which of course leads to fights
To see who is in charge

Young males will try to take over
With all of their strength for their rights
They snarl and they growl
They bark and they howl
And as teeth are their only weapon
They rip out huge chunks with their bites

SNARL!

GRUNTING!

A male wild boar has unusual teeth
That keep growing all of his life
They get powerful and big
And he uses them to dig
Like a sort of very handy
Hunting knife

But don't confuse him with a pig!

A male wild boar charges
Grunting to a fight
Using all his fearsome might
He clobbers with his heavy head
And slobbers from his jaw
He slams and pounds
Goes many rounds
Deep into the night

But remember what I tell you
It'll serve you very well...

Don't confuse him with a pig!

Little lions learn to fight
When they're still very young
Trying out all manner of moves
In what looks like bunches of fun

But during a play-fighting rumble
As they tumble and stumble and fumble
They may mumble and grumble
It may look like a jumble
But they learn to be humble
They learn not to crumble
Until all of their lessons are done

RIP!

POW!

Don't these 'roos
Look like they're boxing?
Ducking and dodging
Each other outfoxing?
Yes, they do
But they're not
No, they're really not boxing.

What they have to do
So that they can fight
Is make a huge effort
To keep standing upright
This means that they balance
On their long, strong back legs
And with their short front ones
Using all of their might
By pushing and shoving
And locking in tight
By jabbing and grabbing
And thrusting and stabbing
And being very quick
They might just be able
To lash out and kick
Wow! POW!

Foxes always hunt alone
And do not care to share
Whatever food they catch
They're awfully quick
To pick a scrap
With any sneaky, cheeky chap
Who tries to nick their patch

Foxes fighting for living space
Stand on their back legs, face to face
Front paws on the other guy's shoulder
At first they'll snarl, then nip and bite
And as the fight gets more uptight
It's a race to push him over

SNARL, NIP SHOVE!

The point of it all
Is to get him to fall
It's the fox who's left upright
Who wins

STOP, THIEF!

Does the egret have a secret?
Something we don't know about?
Well, he cannot really tell us
He can't really shout it out

It could be the egret's secret
Is the way he stands so still
Hardly moving his long long legs
Until with his strong sharp bill
He PRONGS it in the water
And pegs himself a kill

Perhaps the egret's secret
And it's not so very nice
Is that he's very quick to steal
Some other egret's meal?
Oh yes indeed
He'll swiftly wheel
And nick it in a trice
And yes indeed
You've got that right
It ends in quite a mighty fight!

These brightly-colored birds
Just as their name implies
Live by eating mainly bees
And anything little that flies

They hunt early in the morning
And just before the night
Darting, dipping, diving
Catching their prey in mid-flight
The bird will return to its perch
With its catch
For first it must take out the sting
But some other bird
Robber bird
Searching for food
Might SNATCH!
Forcing a fight on the wing

Food is not the only reason
For a fight
The long migrating season
Comes before the mating season
Oh the fights
For the rights
To the best of nesting sites!

SNATCH!

A cuttlefish is not a fish
He's related to the snail
There's a fin all round his body
And he hasn't any tail

The males will fight for ladies
Putting on a great display
Changing color, charging, biting
Until the loser's chased away

CHASE!

Jawfish dig a burrow in the sand
Where they spend a lot of time at home
Alone
They really cannot stand
To let a new boy on their land
So they will never, ever rest
As they do their very best
To drive that pesky guy away
Without delay!

First they'll steal each stone
As he tries to build his home
Zapping at him
Snapping at him
"Get out of my zone!"

If that doesn't do the trick
They'll fly at him real quick
They'll lock their mouths on tight
In a knock-down, drag-out fight!

GO AWAY!

CRAB FIGHTS

A crab has goggly eyes out on stalks
He jogs along sideways
When he walks

A crab has two claws
One is usually a whopper
With this pincer
Our mister
If involved in a fight
Grips the other guy
Pinchingly tight

PINCH!

Lizards can do some very strange things
They'll regrow a tail if they lose it
Lizards will outgrow and change their skins
And though they don't always choose it
They'll fight
Yes, that's right
And it's not a pretty sight
To see a lizard
Another one bite!

CHOMP!

SQUABBLE!

Stag beetles' jaws look like antlers on deer
But let me make it quite clear
You have nothing to fear

Those jaws might look scary
But you needn't be wary
They aren't used for biting
Only for fighting

Male beetles will wrestle
And squabble over food
They fight over females
They fight over space
With those freaky looking
Sneaky looking
Mouthparts on their face

WRESTLE!

STAB! JAB! GRAB!

Did you know
That a wasp can go
Many hours in a fight?

It was news to me too
But it seems that they do
To prove which one is right

They'll fight over a nest
Over which mate is best
And they don't only just bite

They'll tear off wings
Even kill with their stings
It's a hideous, horrible sight!

STING!

MORE ANIMAL FIGHTS

PELICANS

RAVENS

TURTLES

IGUANAS

HORSES

DEER

TURKEYS

BATS

BUFFALOS

LIONFISH

ANTS

Acknowledgments

Photography by: Nicu Baicu, Jeff Banke, Nick Biemans, BlueOrange Studio, Rich Carey, John Carnemolla, Arto Hakola, Jiri Haurelju, Eric Isselée, Sándor Kárpisz, Konstantin Kikvidze, Mitja Kosir, Herbert Kratky, Keith Levit, Chet Mitchell, János Németh, Cloudia Newland, Pavel Plotnic , Evgeny Prokofyev, Cheryl Ann Quigley, Jennifer Sekerka, Johan Swanepoel, David Thyberg, Teguh Tirtaputra, Mogens Trolle, Tom Viggars, and Worlds Wildlife Wonders.